The Pygmy Goat
a practical guide to keeping and breeding

by
Doreen Wright

Spindrift
Print & Publishing
1996

© Copyright Doreen Wright 1996

ISBN: 1 898762 07 4

*All sketches and drawings used in this book
are the work of*
Betty Hesse

*All photographs (unless otherwise stated)
are the work of*
Anna Oakford

Spindrift Print & Publishing
Second Marsh Road
Walsoken
Wisbech PE14 7AB

Mum

with love

Acknowledgments

My thanks go to my long suffering and mainly uncomplaining husband for his support throughout the writing of this book. Thanks also to those who have nagged me to "get it down on paper" and given me encouragement.

CONTENTS

	Page
Introduction	1
Chapter One	
Description	2
Chapter Two	
Horns	8
Chapter Three	
The Pygmy Nanny	14
Chapter Four	
The Pygmy Male	16
Chapter Five	
The Wether	20
Chapter Six	
Housing	22
Chapter Seven	
Feeding	26
Chapter Eight	
Health	29
Chapter Nine	
Handling	33
Chapter Ten	
Breeding	37
Chapter Eleven	
Arranged Colours	43
Chapter Twelve	
Kid Rearing	44
Useful Information	49
Index	50

All measurements in this book are Imperial, a conversion table is shown below.

Imperial	Metric
1 inch	2.54 cm
1 foot	30.48 cm
1 fl oz	28.4 ml
1 oz	28.35 g

To find the approximate weight of your pygmy goat measure round the heart girth, just behind the forelegs as illustrated and use the chart below.

Inches	lbs.	Inches	lbs.	Inches	lbs.
10¼	4½	17¼	19	24¼	51
10¾	5	17¾	21	24¾	54
11¼	5½	18¼	23	25¼	57
11¾	6	18¾	25	25¾	60
12¼	6½	19¼	27	26¼	63
12¾	7	19¾	29	26¾	66
13¼	8	20¼	31	27¼	69
13¾	9	20¾	33	27¾	72
14¼	10	21¼	35	28¼	75
14¾	11	21¾	37	28¾	78
15¼	12	22¼	39	29¼	81
15¾	13	22¾	42	29¾	84
16¼	15	23¼	45	30¼	87
16¾	17	23¾	48	30¾	90

INTRODUCTION

The pygmy goat is becoming increasingly popular, both as a pet and as a show animal. An attractive, charming little being, it is easy to handle and care for and cheap to feed having the added bonus of nanny giving up to two pints of milk a day, if so desired. There are two distinct types of Pygmy goat although in this country today, most are a mixture. They tend to identify more, at least in stature, with those found in West Africa. The pygmy is not, as yet, a breed recognised by the B.G.S. (British Goat Society), and therefore cannot be registered with them, however, many local goat clubs are now including Pygmy Goat classes in their shows.

Although the pygmy is little trouble, its needs requiring a modicum of time and effort, it must be borne in mind that each and every day of the year, without fail, it has to be attended to, and this can prove to be rather less than exciting when you find yourself having to brave the elements in the depths of winter! This includes the billy who is entitled to, and deserves, the same care and attention as that given to the nanny, his health and welfare are equally important and in the interests of us all. Let's face it, when all is said and done, without our studs there would be no procreation of the species.

I, for one, have never regretted the years spent as a keeper of this delightful goat and I hope to savour the pleasures of being associated with them for many more years to come.

Chapter ONE   Description

The pygmy, or dwarf, goat originates from Africa where it is still kept, mostly in desert and scrub areas and is bred mainly for meat. It is thought to be an early ancestral form of the larger domestic goat which has been kept for centuries in Africa and Asia. These small goats are found over a large area of equatorial Africa ranging from the mouth of the Senegal through Central Africa to Southern Sudan. There are two very distinct types, one having disproportionately short legs with plump body and broad head, the other being a slender, more normally proportioned goat.

These are the two extremes, if you like, the first being typical of the West African dwarf goat and the other being found in Southern Sudan, with, as you move across the regions between these two areas, many indeterminate types known variously as the "Egyptian", the "Himalayan" and so on, depending on the region in which they are found.

It seems clear that the one we know as the "Cameroon" (or "Blue") is typical of those found in the Guinea Zone of West Africa and might, therefore, be better named the "Guinea" Goat. At the other end of the scale is the slender so-called "Nigerian" which could easily be compared to a miniature dairybreed goat and relates more to Southern Sudan, hence, "Sudanese" would perhaps be a more appropriate name. However, because these two types have been commonly known as the "Cameroon" and the "Nigerian" since their introduction into this country more than 25 years ago they will be referred to by these names for the purposes of this book. These two extremes are quite different in general appearance and horn structure, the overall impression of the Cameroon being one of a grey, squat, chunky little goat and by comparison, that of the Nigerian, a sleek, leggy, finer boned goat of varied colouring.

The Cameroon (or Blue) from West Africa, is traditionally always predominantly grey (from silver through to dark grey) with a black mask, black socks, a white star or flash on the forehead and ears speckled with white. The head is broad and the body heavy, being wide across the back and extremely full barrelled and one could be forgiven for thinking the animal positively overweight. The coat is long and continues thickly down short, sturdy legs with quite heavy little hooves. The horns of the male are set sideways to the head, then rise up quite straight and at the tips either curve inwards to face each other, thus forming an arch above the face, or flick outwards at the tips.

Perhaps a better description would be to say that they seem to have been given a quarter turn at the roots so that the bevelled front of the horns are placed sideways to the head giving a view of the side edges of the horns as you stand face to face with the goat. Those of the female rise up from the forehead, some curving in quite a wide arc towards the face, others with just a curl at the tips but again, going forwards towards the face, giving the impression in both cases that they've been placed on the head back to front.

The Nigerian is not nearly as heavy bodied as the Cameroon, is longer legged, short coated and comes in various colours. It can be mainly white with markings of black, or tonings of brown, from a creamy beige through to auburn to dark chocolate brown, or the reverse way with white markings on the darker background. Or it can be tri-coloured, a mixture of any of these colours. It should never be all white, nor should it have any grey areas, not to be confused with a mixture of black and white, or brown and white hairs, giving a powdered effect which might give a grey appearance. The coat is short and sleek, legs being very slim with dainty little hooves. The horns of the male rise up from the forehead with a sweeping curve backwards, the tips sometimes turning outwards to left and right of the head. Those of the female are similar but less substantial.

All horns do, of course deviate to an extent, some perhaps, have a more pronounced curve or are longer than others or more substantial. But my observations bring me to the conclusion that the common factor in the male Cameroon is that the *sides* of the horns are always displayed either side of the head rather than the *front* of the horns, whatever course they take at the tips. Those of the female Cameroon always curve or curl to some degree *forward* rather than *backwards*. Nigerian horns in both male and female, always sweep and arc to varying degrees, backwards away from the face.

I'm bound to say that there are obviously, naturally occurring mixtures of these two extremes. For instance, the Cameroon crosses over into bordering regions producing offspring not completely true to its type, and so on, from border to border. I wonder if perhaps the indeterminate types found between West Africa and Southern Sudan are, in fact, merely the results of the wanderings and meeting up of the two extremes over the centuries. However, the two extremes are clearly evident in this country too, though certainly not in the numbers they were many years ago when I was first introduced to the pygmy goat. I'm afraid the indeterminates have well and truly taken over, mainly due to either unintentional crossing because of the breeder being unaware of the existence of the two clear varieties and as a consequence taking nanny to "Joe Bloggs down the road" who happens to have a pygmy billy (whatever it's like), or deliberate crossing in an attempt to achieve a goat with the chunkier look and shorter legs of the Cameroon but boasting the varied colours of the Nigerian.

The results of these experiments are, to my mind, very often not at all pleasing to the eye, having seen time and time again very odd combinations of characteristics. For instance, a heavy bodied goat (which is true to the Cameroon) perched on the longer, slender legs of the Nigerian, having a short coat, also from the Nigerian, with the colouring being mainly grey, from the Cameroon, but with patches of brown from the Nigerian. If it has been disbudded it's anyone's guess what shape the horns might have been - perhaps one of each from the two types!

I do feel that if the breeder has one or both of the extremes they should be allowed to retain their true characteristics by pairing up the female only with its true male counterpart if possible, otherwise I fear these two very distinct types will become non existent. I'm afraid the stage has already been reached whereby the task of finding fresh blood which one can be sure of, is proving well nigh impossible.

Unfortunately, because standards laid down at many shows lean far more towards the build of the stockier Cameroon, the correctly leggy, finer boned Nigerian is usually dismissed out of hand. So yet again, the breeder with a Nigerian type, anxious to gain recognition, will resort to using a Cameroon billy in order that the offspring will, hopefully, be shorter legged and stockier and so improve their chances of gaining a place at championship level.

Ways of breeding back to the original with crosses in order to increase the numbers of the two extremes as well as introducing fresh blood are being tried. However, great care should be taken as it could be only too easy to lose sight of the correct appearance of these two extremes as subtle differences increasingly occur, over a period of time thereby altering the look of the goats, but nevertheless, being accepted as true to their variety. In reality, of course, the goats could bear only a resemblance to the original and what a pity that would be, though it's probably fair to say that even our present day "extremes" have changed somewhat in appearance over the centuries.

I'm no expert in these matters but it's said that once the two extremes have been crossed it's virtually impossible to eradicate characteristics from each of the types occurring in future generations. These inconsistencies may not be immediately apparent in kids but could develop with maturity. I'm told the genes of both types will always be lurking in the bloodline waiting to put in an appearance. So if finding fresh blood that you can be sure of is impossible and you're anxious to keep your stock good, rather than using a suspect stud or one that looks right but cannot be checked through lack of history, you might find it preferable, as has been suggested, to practise in-breeding and if by so doing, really bad faults crop up occasionally, you must cull. This would seem to be the only sure way to guarantee that the kids of your true Cameroon or Nigerian will never throw out signs of the alternative in their offspring. If you're certain you're the owner of either of the two extremes and are concerned about the lack of fresh blood, this could be food for thought and an option worth considering.

I have to say that in their wild state in Africa, pygmy goats are not at all like the well cared for (generally speaking, that is), properly fed, attractive little animals we see in this country. A man who called at our home some years ago was most interested in my small herd of pygmy goats. He had recently retired and moved back to England from Africa where he'd lived and worked for many years. He said my goats were hardly recognisable as the pygmy goats he had been used to seeing in Africa. They were quite a common sight there, poor, scraggy little animals with rough coats, ceaselessly foraging for almost non existent sustenance, hardly worth a second glance. Even the small herd owned by the odd native fared little better, obviously undernourished and scouring badly with undoubtedly nothing done about it, they were a miserable sight to behold. What a difference care, attention and regular feeding can make, he said. It certainly doesn't seem to be in the nature of the pygmy goat to be fastidious in its eating habits and without doubt the reason for this, and its inherent hardiness, stems from its fight for survival in its country of origin.

The pygmy, whether an indeterminate or one of the two distinct types, has a wider girth in relation to height compared to other breeds. With a full barrelled body, its bulk, particularly in the case of the Cameroon, suggests a permanent state of advanced pregnancy which is enhanced by the legs and neck being short in relation to body length. Dish faced with prick ears it comes in various colours and markings. Heights range from 16" to 23" to the shoulder for nannies and possibly up to 24" for billies and on average can be compared to a medium sized dog.

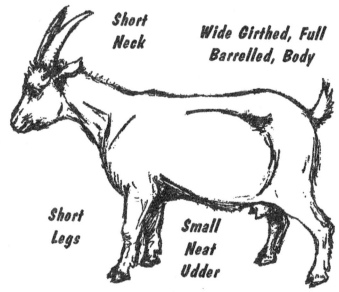

Both male and female can have "toggles" or not. These small furry appendages that hang down from either side of the neck do not serve any purpose, they're purely decorative. They will be present, or not, from birth.

In maturity, both male and female have beards, the males' being dense, the females' sparse. When I take my nannies to breed shows I never cease to be momentarily lost for words when occasionally some people refer to them in the masculine because of their beards, despite their nice little rounded udders with miniature teats proclaiming their femininity. At championship shows the beards of the females are usually trimmed off so perhaps some folk have only seen them bereft of their whiskers and not as they are naturally, complete with wispy beard.

The female is considered fully mature by the age of two years and the male at two and a half years. It is very hardy and will withstand extreme cold without complaint, however, like its larger relation, it doesn't like rain, especially when coupled with wind and will flee for shelter at the first signs of an impending shower, although I'm sure I cannot be the only one to have a little nanny who, when she's in season, will happily put up with rain in order to stand for hours pressed up against the fence, totally engrossed in her fascination with billy, equally enamoured, parading up and down on the other side.

My goathouses open out on to a concrete area and paddock, therefore, in order to keep them hardy, I'm able to leave the doors open during daytime throughout the year so that they can wander in and out as they please whatever the weather conditions. Only rarely do I find it necessary to shut them in, and this is when wind is blowing rain or snow through the open doors making their bedding damp. And it should be remembered that winter is the time of year when most breeding females are likely to be in kid and throughout the five month period of gestation it's essential for them to have exercise. By allowing my goats complete freedom of movement, even during winter months, the foundations are laid for trouble free kidding in the spring.

Pygmy goats are grazers as much as they are browsers and will meticulously go over even a freshly mown lawn, although having said that, don't expect them to replace your lawnmower. Their painstaking efforts in search of food in their natural environment of barren desert land seems to have provided them with an in-bred instinct to find and obtain sustenance from the most meagre of plants.

Chapter TWO   **Horns**

The question of whether to disbud a pygmy goat has often arisen over the years. I have to say all of my goats are horned simply because I was given no choice and that's how they were when I purchased my first breeding pair some nineteen years ago and as it's not advisable to mix disbudded goats with horned ones, subsequent additions to my herd have been allowed to retain their horns. Whilst admitting to a personal preference for the pygmy goat complete with horns as nature intended, there are many who prefer the look of an uncluttered head and disagree that with their removal some of the character of the animal is lost. But aside from appearance there's often doubt that horns could be potentially dangerous to the handler. For me, the very smallness of the pygmy takes away any apprehension I might feel should I be confronted with a restless horned animal of larger dimensions. However, if I had a milking goat I think I might prefer it if the horns weren't there for conceivably the goat could suddenly turn its head whilst being milked and unintentionally cause injury to the head or face area with a horn, and I would be wary of a small child mixing freely and unsupervised with horned goats, for in the case of the pygmy the horns could be on a level with the child's head, another potential hazard. But aside from these two examples of possible danger, I can think of no other reason for depriving a pygmy goat of its horns, if you'd rather not.

It seems to be a foregone conclusion with some who have no knowledge or dealings with goats, that they're an animal whose nature it is to butt for no good reason. This is possibly reinforced by the age old cartoon-like drawings with which most of us are familiar, depicting the classic situation of a goat, head lowered, topped with formidable looking horns in hot pursuit of a fleeing human being, whose hands clutch the posterior protectively in fearful anticipation of a butt guaranteed to send the hapless victim unceremoniously into the air. But this portrayal of the goat is purely mythical. It's generally accepted that, in fact, they're naturally gentle creatures responding readily to kindness and affection and it's considered that the rare goat that persists in butting a human being must have formed the habit through provocation at some stage in its life and as a consequence, will automatically defend itself in this manner at the mere sight of anything on two legs.

A way to prevent this aggressive behaviour would be to stand face to face with the offender, catch hold of the horns and push slowly backwards into a corner. Hold for a few seconds, then release and having proved your superiority the goat will quietly walk away. This procedure may have to be repeated from time to time as a reminder that you are in control.

Top:
Male Cameroon type Horns are narrower from the front and can point either inwards or outwards at the tips

Bottom:
Male Nigerian type Horns are broader from the front.

Pygmy goats do use their horns to establish leadership within the herd and periodical altercations between them will be resolved by a set routine in the form of horn clashing. I've never known injury caused, or indeed, intended by this activity, it seems to be simply a matter of proving superior strength and stamina. They confront one another, rear up on hind legs and on the way down, clash horns with incredible precision. This is repeated for as long as it takes (usually no more than a minute or two), for one goat to walk away, thereby conceding victory and superiority to the other. I am, of course, referring to nannies and/

or wethers. A similar confrontation between entire males would, I imagine, be a different proposition altogether and I've certainly never felt inclined to put my studs to the test.

Young kids with barely visible buds will imitate their elders by re-enacting this horn clashing ritual in play with one another and a lone kid will practise on an unsuspecting, smaller than itself, victim. Many a time I've watched with amusement a macho kid stalking a minding-its-own-business hen, meandering about totally engrossed in pecking at the ground, only to find itself rudely interrupted by a well aimed head butt to the rear, sufficient to send it rapidly on its way, feathers flying, noisily clucking and squawking in protest. Meanwhile, kid swaggers off, smugly pleased with his prowess and no doubt content in the knowledge that he's proved himself just as capable as his elders in putting paid to an imaginary adversary. But nothing more serious than loss of dignity is suffered by the hen and peace is restored as pecking is resumed with increased urgency as if to make up for lost time.

Horn clashing as described is quite normal and harmless between horned goats but it would be a different cup of tea if practised between a horned goat and a disbudded one. So I would like to emphasize, as I mentioned earlier, that disbudded goats should not be allowed to mix freely with horned ones. The disbudded goat even if an established member of a horned herd, would be at a great disadvantage when confronted periodically by horned goats intent on establishing pecking order and could be quite badly injured during the process. And I don't know if other breeds of goats are similar in their attitude towards a newcomer to the herd, but pygmies can be little bullies. To them, a stranger needs to be put in its place without delay and even if the newcomer is, like them, horned, a little supervision wouldn't come amiss when introducing it to its peers and it will soon be accepted into the herd. But if it's a disbudded goat being introduced into a horned herd, the situation could be fraught with danger.

A man visited me a few years ago with a disbudded nanny to be mated with one of my studs. He'd acquired her from a lady who'd purchased this hornless nanny and put it in with her horned goats without realising the potential danger. The little nanny lost an eye during the brief fracas that ensued and although it was months after the accident that she was brought to me for servicing by my stud, she was so distressed at the mere sight of my nannies grazing peacefully in their paddock that I had to advise the man to take her home without attempting to mate her and suggested he tried again the following year. It's really not worth taking a chance when this sort of accident can so easily happen.

Although smaller in size, my goats were quite capable of putting our Airedale terrier in his place when the need arose. I use the past tense because, sadly, he's now departed having lived a long and happy life. "Tom" was his name and he was introduced to the goats, on a leash at first, from day one and played many a game of tag with them to their mutual enjoyment.

However, as an over exuberant youngster he sometimes outplayed his welcome making a general nuisance of himself and the goat on the receiving end dealt most competently with this situation. Refusing to be cajoled into further play, she'd revolve slowly on the spot with eyes riveted on Tom while he circled around her in a mad scamper, tail awag, yapping excitedly, every now and again darting towards her as close as he dare, then away again, in an attempt to tease her into chasing him. Completely unruffled she'd stare him out and bide her time with remarkable restraint. Suddenly, when least expected and as quick as lightning, she'd deliver a well aimed butt to his hind quarters, but with only sufficient force to serve as a warning. It made him yelp but more I think, in surprise than anything else and he'd beat a hasty retreat, tail between legs, until the next time. As Tom matured, he didn't need this sharp reminder to restrain his unruly playfulness and seemed to sense when enough was enough so would give in with good grace and simply join the goats in their peaceful wanderings as if he was one of the herd. They became so confident in his friendship that he was even allowed to visit and nuzzle their newborn kids within an hour of kidding.

I'm bound to say that I am wary of my stud billies. For most of the year, they're placid creatures but in the autumn, when the mating instinct is at its strongest they might demonstrate their annoyance in no uncertain manner, despite their smallness, should you stand in the way of the object of their desires, i.e. a moon eyed, in-season nanny posing invitingly against the fence of his paddock. I learned my lesson many years ago when, all too casually, I entered the paddock of one of my billies to change his bucket of water and unwittingly obscured his view of a particularly desirable nanny wickedly flaunting her assets in a most provocative way. Quick as a flash, he took a sideways swipe at my leg with his horns to get me out of the way, and did I move! Fortunately, I was wearing wellies over thick socks and jeans so a bruised leg was all I had to show for my carelessness and I've made sure that I've not provided the opportunity for this to happen again by treating him with a little more respect.

Lest you've formed the opinion that my goats are continuously engaged in skirmishes of one kind or another, let me assure you that ninetyfive percent of their time is spent quietly grazing or indulging in play, climbing on and jumping off various simple wooden structures my husband has made and placed about the paddock for their enjoyment, or contentedly cud chewing whilst watching with steady unflinching gaze the antics of others and apart from the isolated incident with the stud (and that was my own fault), I've never been on the receiving end of any aggression.

The disbudding of a pygmy goat isn't quite such a simple affair that it is with a larger animal. It's out of the question to disbud at a few days old using an injection into the head as is done with its larger relation. The head is too small at this early age, leaving very little margin for error, therefore, the slightest miscalculation could have dire consequences.

Some vets do disbud at about four days old, or when the buds can be felt but are not visible to the eye but an inhalation anaesthetic is used. Others, including my own vet, prefer to use the traditional method of injecting into the head, but not until the buds have grown sufficiently in height and circumference as can be likened to a miniature thimble, when, at such time the head will have broadened considerably. The spread of the roots that can be felt around the base of the buds is also taken into account. The time is right for a billy, who matures earlier than a nanny, at 2½ to 3 weeks of age and a nanny, at 4½ to 5 weeks. If it's left any later, disbudding probably cannot be carried out at all.

I can only speak from my own experiences of the latter method and say that in nineteen years no problems have arisen and the kid suffers no ill effects during or after disbudding, indeed, in the time it takes to arrive back home with kid cuddled on my lap, it's on its feet and running towards its mother for a drink. An advantage in this method is that buyers of my kids have time, 2½ to 4½ weeks, depending on whether their choice is a billy or a nanny to decide if they wish their kid to be disbudded or not.

Although rare, regrowth of one or both horns (often in a distorted fashion) might occur after disbudding in which case a return visit to your vet will be necessary.

Chapter THREE   **The Pygmy Nanny**

Originating as she does from an equatorial climate, the female's pattern of breeding allows her to be capable of producing kids twice yearly, but in order to keep her fit and strong, producing healthy kids and to enjoy not only a long life but one of good quality, it's considered unwise and unfair to encourage twice yearly kiddings. From the age of about five months, oestrus occurs for approximately three days out of every three weeks throughout the year, the strongest indications being in the autumn. Although obvious signs (to our human eye) are lacking throughout the summer, these will become evident from September, gradually increasing in strength by November. Period of gestation, as previously mentioned, is five months.

As a milker, she's capable of giving about two pints daily - not a lot, but often sufficient for the personal needs of anyone whose requirements do not warrant gallons of milk each day. She'll not produce milk without breeding and will need to be bred every year in order to keep her in milk. She'll dry herself out, no assistance is necessary, she's unlikely to suffer from mastitis.

However, if you don't intend using her as a milker, the kids are best left on her until they're ready to go to their new homes. This should not be before they're twelve weeks old, when, although they'll still be suckling from Mum occasionally (when she allows it), they can be considered completely weaned having been eating concentrates and hay, and grazing and drinking water out of the bucket for some weeks in between feeding from Mum. For the first few days after the kids have gone, her udder, often referred to as her "bag", will grow very large and taut until it looks fit to burst, but don't worry, she certainly won't, in fact she'll probably not take kindly to any attempts you might make to ease off some of her milk. Just leave her alone and within a few days there will be a gradual reduction in its size and tautness until it's back to normal proportions.

Most pygmies seem to be kept purely as pets, not for breeding, either for the purpose of milking, or, as in my case, for the pleasure of

Dennis and Doreen with pygmy kids ~ the branch normally forms part of a favourite play area.

A Nigerian type pygmy female suns herself

the kids which are sold off at twelve weeks of age. But for anyone intending to breed, for whatever purpose, I would advise avoiding the goat at the smaller end of the scale, for these seem to be the ones that are likely to experience kidding problems, in any case, these are best left until their second year before mating.

There are some breeders who are singling out what can only be described as runts, from which to breed, in an attempt to bring down the size even further. These are poor little scraps totally unsuitable for breeding, in fact, usually do not ever experience oestrus, but if they are capable of conceiving they'll most probably abort or experience severe kidding problems resulting in dead kids, in fact, are lucky if they manage to survive themselves. So the breeder needs to go for the larger females between 18" and 23" for easy, trouble free kidding, indeed, the pygmy is renowned for its hardiness in every sense of the word and apart from the occasional exception, will kid easily without help. However, there is always the exception and in this case one must remember that her smallness does not allow entry into her with your hand as is possible with the dairybreed. A couple of fingers is usually all that can be managed and if that cannot put matters to rights your vet will probably opt for a caesarian. But this situation is rare in a healthy, average sized nanny.

Single kids or twins are the norm for pygmy goats with triplets being fairly unusual, however, I did hear of a Nigerian type having five kids, two were mummified but I understand the other three survived.

Chapter FOUR   The Male

Due to his equatorial origins the pygmy billy has no set period of rut and from the age of about four months will perform his marital duties at any time of the year whenever nanny is in season, although his mating instincts too, are strongest in the autumn. Barely discernable during summer months, the aroma associated with a billy begins to strengthen from August onwards, reaching a peak in November and remaining so throughout the winter. He has a far less pronounced unpleasant odour than his larger relation and this is thought to be due to his smallness coupled with his lack of a set rut period.

His fencing and house should be strong for despite his diminutive size a flimsy shed or fence would soon be demolished if the mood took him. A state of mutual respect between him and his keeper should be nurtured. It's no use taking on a stud goat if you're nervous of him, he'll sense it and soon take advantage of the situation.

It should be remembered that however sweet and cuddly he is as a kid, he's going to grow up knowing he has a job of work to do and in no way then, can he be treated as a pet. That isn't to say, of course, that good handling as a youngster won't help with tasks that have to be carried out not only when he's a kid but also when he's mature such as worming, hoof trimming and so on, never mind the daily tasks of feeding, watering etc..

One must be prepared to accept his courting habits which leave much to be desired, having, amongst other things, a penchant for spraying urine over his face and chest which adds to the odour produced by his scent gland. Together with other similarly, "not very nice" behaviour, his lip will curl and tongue loll out and he'll serenade nanny in a loud, bellowing, gruff voice, often reminiscent of a yodelling bull in full throttle. Eyes roll heavenwards, nose quivers emitting snorts and sneezes, all this accompanied by impatient scraping of the ground with a hoof. All this is natural in a healthy virile stud and is greatly appreciated by a nanny who will find everything about him utterly irresistible and succumb only too readily to his advances.

One of my billies sprays himself to such a degree I wonder where it all comes from. His complete abandonment at these times results in him frequently missing his target. After having been on the receiving end of one such incident, I've since made absolutely sure I've not been within striking distance. His legs become wet to the point of saturation and have to be greased occasionally with vaseline to prevent soreness.

My studs are happiest when they can see the nannies at all times during the day, therefore, paddocks for each are set alongside each other divided by a strong fence.

The pygmy has become a preferable choice of billy for some dairybreed goatkeepers who require the services of a male simply to keep their nannies in milk. He's cheaper to feed, his smallness makes him more easily manageable and his aroma is less offensive to sensitive nostrils. However, the discrepancy in size between little billy and large nanny often makes it necessary to provide some sort of assistance so that he can reach the nannies nether regions.

I visited a lady some years ago who kept dairybreeds as well as pygmy goats and used a pygmy billy for both. Glancing from the tall dairybreeds in one paddock to the little pygmies in another, which included the billy, I couldn't help but wonder how he coped with the dairybreeds. Apart from trying to visualize the nannies doing a knees-full-bend in order to bring themselves down to billies level, I couldn't see how it could be managed.

We were standing in a small area between the two paddocks and I'd noticed a wooden structure about two feet high and vaguely assumed it was used by the lady as a stool for certain jobs. She told my husband and I to stand back, opened the gate of the pygmies paddock, called out "COME ON FRED" and billy shot out like a scalded cat, made a beeline for the wooden platform and with a spectacular leap, landed on all fours on top. There he stood steady as a rock, head held high, alert, eager, looking across at the dairybreeds as if to say "Here I am girls, who's first ?". But I'm afraid he was doomed to disappointment that day and a very reluctant, crestfallen little billy was finally coaxed back into the pygmies paddock, his dreams unfulfilled.

This lady reared the offspring of the dairybreeds for meat, but I've heard, and know of others who, if the kids look like the pygmy father, offer them for sale as a pygmy goat, but of course, as they grow to maturity could reach the lofty heights of their dairybreed mother.

A recent observation which astounded me was when I was introduced to a nanny who, at first glance seemed to be a very nice looking, average sized, normally proportioned goat of Nigerian appearance. However, she obviously had dairybreed blood in her because as she'd matured, although her height was unaffected she'd developed huge, long teats which all but touched the ground. These hung grotesquely from her small pygmy sized udder.

Many years ago, I visited a breeder who kept a herd of feral goats from the Highlands of Scotland. As her goats were kept for fibre production, she had no desire to keep her own stud so took advantage of the services of a pygmy billy, which she had on loan at regular times. She didn't say what happened to surplus stock but I suspect there are an awful lot of pygmies about with Highland blood in them.

This seemed to be confirmed when, later, I travelled to view a reportedly good Cameroon billy kid. To my disappointment his colouring was of an evenly distributed brown and grey. He was too young to tell what shape his horns would be or to what length his coat would grow but for me, his colouring alone was enough to convince me he wasn't what I was looking for. When I remarked on this to the breeder the explanation given was that she'd purchased him from Scotland and the brown was the Highland strain in him!!

Another example of crossbreeding, is a trend which was started in America some time ago and has now reached our shores. This is the practise of crossing an Angora nanny with the pygmy billy, the result being the so-called Pigora, the idea being to try to achieve a curly coated pygmy, a goal that so far, I believe, has been unsuccessful.

Without doubt there are many mixtures on the market and I can only suggest to anyone interested in the purchase of a pygmy kid that they ask to see both parents in order to satisfy themselves that it is the product of two pygmy goats and not one of a pygmy father and a mother of a different breed.

For anyone intending to breed, I would advise to go back as far as possible into the history of the breeding pair on offer to ensure that they're both of pygmy origins. In addition, if the intention is to breed with a specific variety in mind, apart from satisfying yourself that the pair you're looking at conforms to the variety of your choice, do try to check that the ancestry too, is true to its type.

Some breeders have more than one billy and so that they cannot see each other it's as well to fill in the dividing fence or they'll play havoc with it, particularly if one witnesses the serving of a nanny by the other.

I must emphasize that although his mating instincts, like the nanny, are strongest in the autumn, if nannies are made available to him throughout the year, no sooner will they have produced one lot of kids than they'll be in kid again. This applies to dairybreeds too, who will be only too willing to help him perform even without the aid of a platform which merely serves to lessen the ingenuity required. So the only time a stud can run with the females, if so wished, is for the purpose of mating and then during the five months period of gestation. But there are two disadvantages in allowing billy to run with the herd in order to put your nannies in kid. The first is that you will not know kidding dates, the second, that the nannies will have a distinct billy smell because of his constant, close contact with them.

Finally, a word about tethering billy. I remember when a breeder went off on holiday leaving her small herd of pygmy goats in the capable hands of a trusted carer. Kidding was over and small bodies gambolled happily around the paddock. Standing in the nannies paddock, last minute instructions were given to the carer although everything had been gone over a hundred times before. But a final afterthought, perhaps the carer would tether billy for a couple of hours a day "anywhere that needs eating down" - giving a vague wave towards the considerable area of land surrounding the paddock. Then, looking up at the hot sun beating down, "but do make sure he can reach shade". The breeder returned from her break. No problems whatsoever, the carer reported happily. And so it seemed, the kids were thriving, billy and nannies looked content and no wonder. It soon became clear that the nannies were in kid again. The carer had dutifully tethered billy as instructed, in a shady corner, *in the nannies's paddock!*

So be warned, nannies will visit billy if he's tethered and he'll manage very nicely, thank you.

Chapter FIVE   The Wether

The wether (a neutered billy), makes an adorable pet. He is a content and friendly little fellow and castration alone, a very minor operation performed by your vet, completely removes the odour which develops with maturity in a whole billy, although it does make him grow a little larger than he otherwise would have done. Due to his smallness he's not sufficiently developed to enable castration to be carried out until at least two weeks of age. He makes an ideal companion for a single goat whether it be a nanny or another wether.

It's interesting to note that the development of horns in a wether seems to be affected by castration. I've seen several of the Nigerian variety neutered at two weeks of age before the buds began to follow their particular course of growth, whose horns in maturity are unsubstantial and no different to those of its female counterpart. Similarly, those I've seen of the Cameroon wether follow the shape of the females' arcing forwards towards the face.

I had my own wether for many years, I decided to keep him as living proof that his smell, as he matured, was no different to that of a nanny because so many people in those early years were unconvinced.

Percy, as I called him, was quite a character. I remember when he was about a year old, a pet rabbit had died. The hutch stood quite high on 3ft legs and after cleaning it out the front was left open so that it would dry out in the sun. I was suddenly aware that Percy was missing. After prolonged searching and calling, I found him. He'd jumped up into the hutch and there he reclined, lazily chewing the cud, completely unconcerned that I'd been wasting time looking for him.

He was a very helpful little lad and gave his assistance freely whenever a job had to be done, like tipping up tins of screws or nails, scattering them everywhere, or running off with the last piece of string which was about to be tied around branches of hawthorn to hang on the fence. And having fights with bales of loosened straw which were placed to hand ready to lay in the goat houses, the paddock ending up looking as if a hurricane had hit it with straw scattered everywhere. But worst of all, at mucking out time he'd persist in jumping into the wheelbarrow at the precise moment a forkful of muck was being tossed into it. He'd emerge from beneath the muck with discoloured straw dangling from his horns and clinging to his body - a disgusting sight.

He couldn't bear to be ignored when visitors came to view the kids and whilst the nannies hovered anxiously but quietly while their kids were being cuddled and admired, Percy did everything he could to transfer attention from the kids to himself - nudging at backs of knees causing a state of semi-collapse to the recipient, tugging at coat sleeves, untying shoe laces and by rearing up in front of the visitor and balancing on his hind legs with front legs against their waist, or thereabouts, forced them to walk slowly and unsteadily backwards providing the onlooker with a unique demonstration of a kind of slow foxtrot performed by a goat with an unwilling human partner.

On one occasion he was making such a nuisance of himself in front of visitors that my husband led him away and shut him in his house so that I could chat quietly and show off the kids without Percy's interference. I was saying what lovely placid pets wethers make and the words were barely out of my mouth when there was an almighty crash. Percy, in his annoyance and frustration at being shut away from all the fun had given a tremendous biff to the door (it would have to be the one with loose hinges) and there he stood on top of the flattened door as if to say "here I am again folks", and I swear there was a triumphant grin on his face. The visitors were startled, to say the least (so was I!), however, to my surprise they left quite happily having ordered two wethers. But apart from his lapses of mischief, greatly encouraged by people laughing at him, he was an endearing little fellow and great fun.

Chapter SIX   Housing

Housing   If you have no readily available outhouses and have to buy or build accommodation for your goats, a sturdy, rainproof, ventilated shed approx. 8ft by 6ft is adequate for two fully grown pygmy goats (or a large dog's kennel is ideal for one but I'll return to this later). The height of the house is optional, anything upwards of 4 ft. will do for the goat but could make mucking out an awkward and backbreaking task unless you're under 4ft. tall! It's useful to leave a space about 2 ft. in, right across the inside of the house at the front before laying straw for the bedding beyond.

The water bucket will stand evenly on this clear area preventing spillages but if you have a playful (naughty!) goat who persists in tipping the bucket over, water will not affect bedding and can easily be swept outside. Having said that, standing the bucket in a tyre will put a stop to any antics of this nature or bucket rings can be bought which are fixed to the wall and hold the bucket off the floor. Another advantage in keeping this area clear of straw is that any abnormal droppings from the goat will easily be identified when sweeping up. A messy tail will show which goat has a tummy upset and prompt action can be taken.

Bedding   To keep straw contained within the area allotted for bedding a plank of wood will be needed about 1" thick by 5" wide and a fraction less in length than the width of the house. This will stand on its edge across the house from wall to wall at the point where the bedding is to begin. To hold it firmly in place and so that it can be removed when mucking out, at the base of the wall fix two upright pieces of wood about 1" square by 4" long, side by side but with a gap a little over 1" between them, on either wall. The plank can then be slotted into and out of the gaps as required. A good layer of straw can be spread over the area behind the plank and the goats' sleeping area is complete.

Some people use woodshavings but if these are of the small finer shavings I would not recommend it for breeding females. It could cling to the teats and cause problems for the suckling kid.

Hay Rack A hay rack is required fixed to the inside of the house with the bottom placed at a height of about 2ft from ground level so that the goat can reach it comfortably. It is not advisable to use a hay net, because horns or legs could become entangled in it causing a nasty accident.

Home made hay rack A simply made structure approx.18" wide by 12" in height by 4" in depth will serve as a hay rack. Made from 4" by 1" wood you will need two pieces 12" long for the two uprights and, one piece 18" long for along the bottom. A piece fixed across the back at the top will be required to screw the rack on to the wall. Tack a piece of 2" gauge wire netting or chain link across the front of the frame. The goat will push its nose through the holes of the wire and tease out the hay. Depending on the number of goats, more than one rack of this size may be necessary or, if preferred a larger rack.

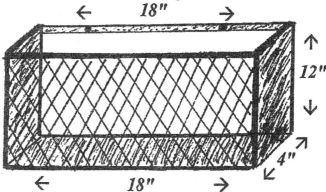

Play area or paddock A fenced area (as much as you can spare), off or around the house. This allows the goat to have freedom of movement and easy access to the house for shelter if it should rain. If goats are housed together in one large shed with free access to the paddock it's as well to have two entrances. I've seen goats attempting to gain entry into a house having one entrance in order to shelter from wind and rain, only to be thwarted by, usually, the leader of the herd lying across the only entrance, effectively denying them access.

They will gather in a forlorn group, plaintively exercising their lungs and miserably contemplating the rain while she languishes across the threshold, gazing at them disdainfully, fully aware of their plight but treating it with complete indifference. Similarly, if you have a shelter of sorts within the paddock because housing is in a separate area, do make sure it has a wide entrance, or leave the whole of the front open so that all goats can enter for shelter without any undue jostling and pushing out of minors.

Any structures i.e. steps up to a sturdy platform or piles of logs placed in the play area or paddock for the goats enjoyment should be sited well away from fencing otherwise these will be used as useful and convenient aids to leaping over the fence.

Fencing   To contain pygmy goats fencing should be 4ft in height but I add another foot for my studs. They're incredibly athletic when being teased by an in-season nanny on the other side of the fence.

Fencing for nannies and wethers paddock   Stakes 3" round and 5ft long, knocked into the ground to a depth of approximately 9" and at 6 ft intervals, then covered with 4 ft high 2" light gauge chain link fencing, stapled to the stakes, will be goatproof. It's worth paying a little extra for treated weatherproof stakes and posts which will delay rotting and so last much longer. I would advise not to be tempted to try and cut costs by using chicken wire or even pig wire. The former will soon be breached and in the long term will cost more in time, labour and money for constant repairs and replacement, than spending out in the first place for chain link. (Here speaketh the voice of personal experience!). Pig wire will be strong enough but I've seen goats push their heads through the holes and because of the presence of horns, cannot withdraw again. Despite agitated struggling, which doesn't improve matters, they're likely to remain in this predicament becoming increasingly distressed until such time someone comes along, notices their plight and with patient manipulation, succeeds in releasing them. For the breeder, another disadvantage with pig wire is that young kids could escape through the holes.

Fencing for billies paddock The billy's fencing needs to be stronger and higher, therefore, sturdier, 6 ft long stakes about 5" round, should be knocked into the ground to a depth of 12" at 4ft intervals. If the billy is disbudded this can be covered, as for the nanny's fencing, with 4 ft high chain link and topped with a narrow rail, at the 5 ft mark. A mere glance at the top rail will convince the billy that his chances of clearing it are next to nothing and not even worth the effort of trying.

Unlike my nannies, who don't attempt to work on chain link with their horns, the billies, whose horns are stronger and more substantial, have been known to amuse themselves in this way, eventually spoiling, if not actually breaking, the fence. I've therefore found post and rail fencing is more suitable for a horned billy. The posts should be 4" square and the rails 3" by 1½", spaced out at about 4" apart up to a height of 4ft, then one more rail along the top of the posts to bring it up to the required 5ft. However, any section of fencing separating him from the nannies paddock will need the lower part, up to about 2ft from ground level, all but filled in to prevent kids getting through to him as they will easily slip through a 4" gap.

Please don't, under any circumstances, use barbed wire for any part of your fencing.

Tethering Do keep an eye on the tethered goat, it could injure itself whilst attempting to gain freedom from an entangled tether or in trying to reach shelter from rain. It is not advisable to tether a goat before the age of nine months.

Storage A dry shed or similar shelter will be required to store hay, straw and a bin in which to keep a bag of goat feed. This must have a tight fitting lid to deter vermin invasion. Hay and straw should be raised off the floor to allow air to circulate, a wooden pallet being ideal for this purpose or planks of wood placed across an edging of bricks will do. It's convenient to have the storage shed sited as near as possible to the goathouses but do make absolutely sure at all times that your goats cannot gain entry into it.

Chapter SEVEN   Feeding

All goats require a varied diet, giving a good balance of vitamins, minerals and proteins. Goats also require a large quantity of coarse fibrous materials if their rumens are to function correctly. Although it is possible to mix your own feed it is usually easier to buy one of the many very good mixes which are on the market. However, it is preferable for the pygmy goat to be fed with a mix as low as possible in protein content.

Goat feed (concentrates) One handful (about 2ozs) in a container night and morning (or 4ozs daily). If you have more than one goat, suggest separate dishes for each placed well apart, otherwise the dominant goat could eat the lot not allowing others to have any at all, never mind their fair share. Goat feed is doubled for the in-kid nanny at approximately 7 weeks before kidding date and I like to increase my stud's ration during his period of working i.e. serving of nannies, although sometimes the billy will become uninterested in food at this time allowing nothing to distract him from his preoccupation with the females.

Hay Very little will be eaten during summer months providing grazing is at the goats' disposal, although it should always be available and during winter months it is essential that the hay rack be regularly replenished. A coarse hay is usually preferred by goats and don't expect it to eat hay that's been on the floor and trampled on or soiled in any way.

Hay can vary in price depending on scarcity and/or the size of the bale. It's best to feel the weight rather than judge by size as a large bale might be loosely packed and have no more in it than a smaller, denser bale. Make sure the hay is good right through by pushing your fingers into the centre of the bale and pulling a little out. It should have a sweet, fresh, pleasant smell. If it smells musty your goat won't eat it. Dusty hay should be avoided for the sake of the health of both you and your goat. Pea, barley and oat straw can be used to supplement hay if this is scarce or to provide variety, but these are not, nutritiously, as good as hay.

Do not overfeed your goat. Obesity can cause health problems and, in the breeding nanny, difficulties in conceiving and kidding.

Water A bucket of fresh water daily, in a corner of the house or paddock. Warm water in winter is not only often appreciated by the goat but will help prevent icing up in freezing temperatures.

Buckets Empty jam and synthetic cream containers made of plastic and with a handle can be begged from your local home bakery. Generally about 12" in height and 28" measured around the lip, these are more than large enough to hold sufficient water for the daily needs of the pygmy goat, serving as ideal water buckets.

Feed dishes Discarded saucepans or frying pans and the like, with the handles removed can be utilized as feeding dishes and are heavy enough not to be tipped over easily. Frying pans are particularly useful when nanny has kidded, these being wide enough to allow kids to push their noses into it along with mother to sample the concentrates.
Note:- Do make sure feed dishes and buckets are kept clean.

The pygmy takes well to a lead and can be taken for walks if wished and by allowing it to browse along the hedgerows it will provide a mixed diet which all goats prefer. Before allowing your goats to browse a hedgerow you must ensure that there are no poisonous plants growing within it. Bearing in mind twenty minutes browsing provides about two hours of cud chewing, this will help reduce feeding costs although at approx. £1.40 per week per goat, feeding is, in any case, inexpensive. For the breeder, the cost will be higher taking into account the in-kid nanny whose intake of both concentrates and hay increases during the later stages of pregnancy and continues throughout the period of suckling her kids. The kids too, will be taking their share of food, starting with small quantities of concentrates from about a week old and building up to a full ration at twelve weeks of age together with taking hay from the rack.

Feeding as described is all you need to keep your pygmy goats healthy and content, but if wished, roots such as sugar beet, fodder beet, carrots, turnips, washed and chopped up, although not essential to their diet, could be provided as a variation when grazing is sparse. Windfall apples too, chopped up and acorns are often enjoyed, but make sure all of these are only titbits, don't overdo it - and they should never be considered a substitute for their basic diet of concentrates and hay. Leaves and stinging nettles, fresh or dried out are popular, as is hawthorn and brambles, these can be bunched and tied to the fence. If you're pruning trees, providing they're not of the poisonous variety, fruit or otherwise, put the branches in the goats' paddock or where it's tethered, it will enjoy nibbling off the leaves and stripping the bark. Rose prunings are also popular, although with all these delicacies what one goat finds palatable another might not.

Grass cuttings Pygmy goats are particularly partial to grass but never give them grass cuttings as the mowing action causes a reaction which heats up the grass and will make them ill, apart from this the cuttings could be contaminated by oil from the mower. Grass from scything or strimming is suitable.

Stripping the bark To prevent your goat stripping the bark of trees, which, although palatable to the goat would eventually kill the tree, paint the trunk with tar to a height of about 3ft from ground level. Goats dislike the smell of tar and will not venture near the tree.

It's very important that you ensure fencing is goatproof and that your goat is tethered securely because if it escapes your garden and/or that of your neighbours, will quickly resemble a desert. In any case, apart from a number of wild plants which are highly poisonous to goats some of the more common garden plants which should be avoided are laburnum, yew, laurel, privet, rhododendrons and clematis, all of which the goat is likely to sample if readily available.

Chapter EIGHT   Health

Over the years, apart from the rare occasion, vet's fees in my experience have been negligible being confined mainly to castration and/or disbudding of kids. However, I'm willing to concede there has probably been an element of luck in this although I do practise good goatkeeping to the best of my ability. I know of some goatkeepers dedicated to the welfare of their goats who have found themselves saddled with hefty bills through no fault of their own. But, generally speaking, providing you start off with a healthy goat, a properly fed, well cared for animal goes a long way towards avoiding vet bills or at least in keeping them to a minimum and will ensure a long, healthy and contented life for your goat. In my experience the average life expectancy seems to be between 10 to 12 years for the male and 12 to 15 years for the female.

There are a few basic routines to be followed on a regular basis.

Vaccination and CAE testing   Your vet will advise you if you wish to take advantage of a vaccination available to protect your goat against enterotoxaemia and tetanus, and blood tests done on a regular basis to confirm that your goat is clear of C.A.E. (Caprice Arthritis Encephalitis).

Worming   Your goat should be wormed at least twice yearly. I worm mine spring and autumn. Your vet will supply worming granules, which can be mixed with the goat feed or a liquid wormer together with a plastic syringe. I alternate between wormers, it's best to do this or the goat could become resistant to the same wormer used on a regular basis. The amount to be used depends on the weight of the goat. How do you gauge the weight of a goat? Well, I find, because of the pygmies' smallness, it's simplest to use my bathroom scales. I weigh myself then, holding the goat in my arms, weigh again, subtract my weight from that registered whilst holding the goat and the difference is the weight of the goat (for another method see opposite page 1).

If you use granules, do examine the feed dishes to make sure all of the tiny white particles have gone, along with the goat feed.

If you use a liquid wormer, the following procedure should be followed. Goats have teeth at the bottom only in front, followed by a gap either side before teeth at top and bottom at the back. Insert thumb of hand into the gap at one side of the mouth, this will make the mouth open. The filled syringe held in the other hand can then be inserted into the mouth on top of the tongue. Slowly spray the liquid over the back of the tongue. If you back the goat into a corner, or tie it to a post on a short rein if you prefer, then stand astride it so that you can lean over above the goats head, you shouldn't experience any difficulty.

Hoof Trimming For your goats comfort, keep an eye on hooves. They will need trimming periodically, especially if running continuously on soft ground. A kid's hooves are best trimmed with sharp scissors having rounded points but as they harden, a pair of straight bladed garden secateurs or suchlike will be suitable and make trimming easy. The young kid can be held in the arms of a helper whilst you trim, carefully avoiding the soft area. A mature goat can be tied to a post on a short rein whilst each hoof is lifted in turn in order to trim. Start with the front hooves so that the goat can see what you're doing and will then not object when you trim the back hooves. Clear dirt from between toes and pare heels to same level as the soles of toes.

However hard you try to look after your goats from time to time problems will arise. Some of the most common signs to look out for are:

Goats excrement The healthy goats droppings are dark brown in colour. They should be firm, dry and pellet shaped, resembling the droppings of rabbits.

Constipation Ash leaves will relieve the constipated goat.

Scouring (diarrhoea) If the goats droppings become soft, indicating the possible onset of scouring, oak leaves in small quantities, are good for binding. But if scouring develops, keep the mature goat shut in its house for at least 24 hours with a plentiful supply of hay and water only. Don't give it anything else at all and this includes goat feed. Once you're satisfied droppings have returned to normal the usual ration of goat feed can be resumed and it can be let out again to graze.

*Cameroon nanny with kids.
Fencing needs to be kid proof and approx. 4 ft. high
for females and wethers and 5 ft. high for males*

*This sturdy structure gives both shelter and a platform on which
to relax or play. It is positioned well clear of the fence to avoid
tempting the goats into attempting to jump out.*

Two week old pygmy kids

Dennis Wright and pygmy mum with triplets
(Photograph courtesy of Eastern Counties Newspapers)

Because a kid will dehydrate quickly, scouring must be treated immediately with a scouring mixture obtainable from your vet and administered in the same way as for worming, with a plastic syringe. This mixture will not be suitable for a mature goat. Contact your vet if there is no improvement after 48 hours.

Lice   Despite the highest standards of cleanliness, your goat could still become infected with lice, these often being picked up from even freshly laid straw bedding. On a dark haired goat the tiny white eggs will be easily detectable around the ears and head, having the appearance of dandruff. Regular examination and the application of a louse powder will control these external parasites. Many well known louse powders are now considered too dangerous to be applied directly on to animals but a powder intended for dogs is suitable. However, take advice from your vet. There are also pour-on liquids applied with a dispenser, but these can be expensive unless you have a large herd.

Seborrhoeic dermatitis   Several years ago it came to light that a type of skin disease was affecting some goats of the pygmy breed.

Unlike zinc responsive dermatitis which is very similar and responds well to zinc therapy, this disorder seems to be resistant to treatment and, sadly, usually ends in the goat having to be put out of its misery.

Of a scaling dermatitis nature it starts around the eyes, ears and lips and between the back passage and genital organs, in other words, sparsely haired areas of the body, but often spreads to other parts of the body too, resulting in permanent loss of hair wherever it strikes. It does not discriminate between the sexes and all ages can be affected. It's been established that it's not infectious and is thought not to be hereditary and although steroids help, they do not cure. There's no evidence of vitamin deficiency being the cause.

As far as can be ascertained, no positive progress seems to have been made towards counteracting the disease, apart, that is, from at least one case known to me where homoeopathic treatment was applied and which seems to have met with success, although as has been pointed out, this should be viewed with some caution due to the fact that the condition is known to be liable to prolonged periods of remission and relapse between re-occurrences of the disease. However, homoeopathy would appear to offer a welcome ray of hope for those with goats suffering from this disorder.

Needless to say, in order to avoid heartbreak, not to mention heavy vet fees, it would, therefore, be wise when buying a kid or mature goat, male or female, to check it over carefully for signs of this skin disease.

Legal requirements

Nuisance. By law all goatkeepers must ensure that their animals do not interfere with the peace and well being of neighbours i.e. excessive noise, invasion of gardens and in the case of a billy, obnoxious smells.

Movement of goats. In order to trace animals should there be an outbreak of infectious disease, new rules regarding movement of animals will be effective from 1996 applying to all goatkeepers and includes the pygmy, even if it's only kept as a pet.

All goatkeepers must register with their local MAFF (Ministry of Agriculture, Fisheries and Food), Animal Health Centre. You will have to give your name and address and will receive a registration form, bearing your herd number. If the goat is for export it will need to be permanently marked for identification i.e. a tag or tattoo. For movements within the U.K. it can be marked in a temporary way i.e. a patch of sheep marker somewhere on the animal, but it must travel with a travel document filled in by the goatkeeper noting date, your address plus that of the destination, the number of goats being transported and details of the identification mark you've chosen for your animals. MAFF will suggest a format but this needn't be adhered to. You must keep records of your herd and these must be retained for three years. Once a year the total number of goats and the number of females over one year or which have given birth, will be recorded.

A book should be kept by the goatkeeper as has been done in the past, noting each occasion their goats leave their land, where it goes and for how long but in addition, noting the animals' identification mark.

Chapter NINE Handling

As I've already mentioned, a large dogs kennel is ideal for a single goat, but I prefer my kids to go in pairs because the goat is a herd animal and is happiest with company of its own kind. However, I've sometimes allowed singles to go. On one occasion a lady wanted company for her pet lamb in the shape of a goat. Pygmies seem to have an affinity with sheep so I was happy to let a little wether go on his own. I often hear from her and sheep and goat graze together and seem to be great company for one another. Other times, singles have gone to dairy-breed goatkeepers, this too works well, the pygmy running happily with its larger relation. And some years ago, a retired lady had one of my wethers. She had most of her garden fenced off with a small shed placed in it which stood only a few feet from her kitchen door.

This little lad lives the life of Riley. He's a regular visitor to her large kitchen-cum-living room where he settles himself down on an old sofa alongside her dog and contentedly watches television. He's taken everywhere with her, in her car for outings into the countryside where he's taken for long walks on a lead - and on most Sunday's he walks with her along the lane to her local pub where he's tethered outside while she has her Sunday lunch. He's well known in the area, in fact, I understand he's become quite a local celebrity.

However, a single goat is generally an unhappy one and unless you have the time and inclination to give it almost constant attention it will bleat loud and long, presenting a most miserable sight.

If you have to part a goat temporarily from the rest of the herd and house it separately, perhaps through sickness or for whatever reason, try to ensure it can still see and hear its contemporaries or it will feel ostracised and demonstrate its unhappiness with loud, incessant, nerve racking bleating, making your life a misery, not to mention that of your neighbours!

When I take stock to breed shows I never leave a goat behind on it's own, at least one other is left too so that it has company. Even so their bleats of protest as we leave, accompanied by those of the goats ensconced in the trailer, have to be heard to be believed, although once on the road they soon settle down. Then there's the homecoming. The goats in the trailer seem to sense the moment we reach the lane leading to our home and become restless, bleating in anticipation and as we turn into the drive the "home alone" goats join in with gusto. What a welcome!

Goats are creatures of habit and appreciate a regular routine, therefore, once this is established, do try to maintain it.

On the whole, they're sweet natured but do have a stubborn streak in their make up which can sometimes present problems. I wonder if I can be the only ignoramus, who, in my early days of goatkeeping had little idea of what to do when a goat steadfastly refused to "walk on"? This can be particularly irksome, not to say embarrassing, if you're out walking the goat on a lead for the sole purpose of allowing it the pleasures of browsing on hitherto untried territory and this kindness is repaid by it coming to an abrupt standstill for no apparent reason. Despite patient coaxing followed by much pushing, pulling and heaving interlaced with dire threats, it refuses to budge an inch. I hasten to add that although a hearty kick to its rear might have crossed my mind at such times and undoubtedly would have done me a power of good, no physical violence was ever inflicted, I promise you.

Your temper, already frayed at the edges, isn't improved if a passer-by happens along and completely misinterprets the scene which, on the face of it, seems to be that of a raving lunatic yelling uncalled for abuse at a poor defenceless animal, whereupon icy glares of disapproval are aimed at you while the goat wallows in the profound sympathy directed at itself. It's nothing short of a revelation to realise how a goat, rising to the occasion, can, in the twinkling of an eye, transform its look of resolute sulkiness and obstinacy into one of pathetic, woebegone, meek humility, giving the observer a totally false impression of a hard-done-by, browbeaten animal, habitually harangued by a sadistic keeper.

However, I did eventually learn that there is a very simple and effective way in which to get a determinedly static goat moving. One thing you must not do is try to drag it along by its lead. This will only serve to make it even more resistant to subsequent efforts to get it going, furthermore, if the collar is pulled tightly putting pressure on the neck in the wrong place, it will collapse. The thing to do is to hold the lead loosely in one hand and give a firm pat to its rear end with the other. This will make it move forward, possibly only a few steps at first, but by repeating this procedure each time it stops it will finally get the message.

Another way in which the goats' obstinacy may manifest itself is when it persistently ignores your commands to "come", so you're left with the formidable task of trying to catch it. A tricky situation this! First I note the following points for you to consider.

1. Never chase it. You might imagine your running prowess and stamina matches that of an Olympic sprinter but I can assure you this will be nothing compared to the speed, agility and evasive skills practised by a goat determined to avoid capture.

2. It's essential to keep your sense of humour intact, even though it might be stretched to the limit.

3. You must be endowed with the patience of Job.

4. You must be as devious as the goat.

Assuming all these fine qualities are yours, I follow on with a step by step account of, arguably, the best way to proceed, the underlying theme being sheer, unadulterated bribery.

You squat down on your haunches keeping perfectly still and with outstretched hand offer a little goat feed or favourite titbit. Meanwhile, in a soft wheedling voice you murmur encouragingly, sweet nothings, such as "you dratted animal, I'll strangle you when I get hold of you", or words to this effect. If you keep at it, the goat, noted for its inquisitive nature, will eventually rise to the bait and slowly approach to see what's on offer. At this point you must resist the urge to sidle awkwardly (being in a squatting position) forwards, to meet the goat halfway as this will only serve to heighten its suspicions making a rapid retreat inevitable. So stay where you are and continue with the soft talk. Sooner or later (generally later), the goat will reach your hand eager to snatch the titbit - and run! In order to prevent this you can do one of two things, the first being somewhat debatable because of gross indignities being suffered by both goat and handler which will hardly make for a lasting relationship. However, it's a method usually favoured by those who have had no previous experience in attempting to capture a goat and is an entirely understandable measure to take given the situation, though I have to say, it rarely works.

You throw yourself at the goat in a rugby tackle as soon as it's within striking distance, but you do risk so upsetting the goat at this unforgivably rough handling that I doubt it will come anywhere near you for a very long time to come, possibly weeks (goats have long memories). Besides, if you bungle this attempt you'd have to start all over again and I may as well tell you here and now you'd stand no chance of success second time round and the distrustful demeanour of the goat as it stares at you accusingly from afar, will be more than enough to convince even the most optimistic of handlers.

So rather than the rugby tackle, a better idea is probably the second method. You hide your frustration, control your temper and give every appearance of being your usual kind, loving, calm self, right up to the split second its nose touches your hand. You then spring into action and in one swift movement, deftly grab collar, a horn, or in the case of a billy, the beard, and with a little luck you should find it possible to hang on grimly whilst you struggle to your feet having completely overbalanced smack into the only muddy patch for miles around, or worse, landed upon nettles, viciously spearheading certain parts of your anatomy, this being particularly painful if you happen to be wearing shorts. But never mind, you've got your goat, that's the main thing!

However, I have to admit to a possibility of this failing too in which case I can only suggest you resort to a third option which is to tell yourself it's no big deal, retire from the field of conflict, have a cup of tea and allow the goat to go where it wants and do what it likes until it becomes bored and decides to return of its own free will!

Collars   For the kid, a nylon puppy collar is ideal. These, of course, are also available in larger sizes and suitable for the adult pygmy goat, or a good quality dog's collar will be long lasting particularly if its made from chrome leather which is weatherproof, although rather expensive.

Chapter TEN   Breeding

I like to arrange for kiddings to take place from April onwards so that kids and Mum benefit from the warmer, brighter weather, therefore, I prepare for matings to take place during November onwards. Come August, I begin to make notes every three weeks of the dates on which nanny shows signs of being in season. This gives me a good idea of the dates in November when I can expect each of them to be ready for mating and will also show the number of days they're likely to be in season. My nannies mostly remain in season for a full three days except as they become old when it lessens to two days, then one day and finally nil.

It's as well to remember that strong mating instincts begin to wane in February and thereafter peter off, the need to mate becoming far less urgent throughout the spring and summer, therefore, a successful service arranged as late as February, is not always so easily achieved.

If you have to travel for stud it's best to choose the day when signs are at their strongest, for instance, in the case of a "three day in season" goat, I would choose to take her on the second day when she's at her most receptive. This will probably save you making return journeys because she hasn't taken or held, and save time wasting for the stud keeper who charges a single stud fee however often you have to return for further matings of the same goat, providing it's within the same season i.e. the same year. If you don't get it right one year and decide to give up until the following year, don't expect a refund of your stud fee or, alternatively, a free service the following year.

Another possible reason for your nanny returning into season after a seemingly satisfactory service could be that she's failed to hold because of the journey, although this is unusual unless it's a long journey, i.e. perhaps two or three hours on the road after mating. It's rarely, if ever, the fault of a proven stud if the nanny fails to take. I've known of billies being literally on their last legs and they'll still put a nanny in kid.

If the male fails to serve the nanny, a stud fee is not expected, but a satisfactory service should be paid for at the time. In my experience, the sign of a good service is when nanny visibly hunches up her body for a few seconds as billy withdraws from her.

Oestrus (in season discharge) occurs every three weeks over a period of approx. three days throughout the year, the strongest indications being in the autumn. It is during the time she discharges when she'll call (bleat) persistently with tail wagging spasmodically, that billy will serve her. Any other time neither will be interested, that is to say, billy might explore optimistically, but nanny will not stand for him and unless she does, he cannot serve her. The discharge is a colourless, sticky substance and particularly when she's a youngster looking forward to her first service, may not be very profuse or noticeable.

Having your own billy on the premises should present no problems in determining whether your nanny is in season or not but otherwise, if you're experiencing difficulty, a "billy rag" could be the answer. You ask the stud keeper to rub an absorbent cloth (perhaps a strip of old towelling) thoroughly over the head, between the horns and over the face and chest. The cloth is then sealed in a screw top jar and taken back to nanny. Several times a day and possibly for a few days, you remove the cloth from the jar and hold it gently over nannies nose. Bringing the odour of billy to her in this way often brings her into season. Providing the cloth is always re-sealed in the jar after use, it can be used repeatedly, on all of your nannies if necessary.

On one occasion when I had to prepare a billy rag for a breeder, I offered to deliver it to her but when I arrived, she was out so the jar, labelled "BILLY RAG" was left in her post box on the gate. This gesture was promptly reciprocated by my being presented with a cloth contained in a jar clearly labelled "DEN'S RAG", (my husband's name) the implications of which didn't bear thinking about - and still doesn't come to that - and in case you're wondering - no! her husband's name wasn't "Bill"!! So enough said and I'll pass on without further comment to the more sober side of the subject.

Another way worth trying is to press the palm of your hand firmly against nannies tail holding it down against her for about twenty seconds or so. This will cause her to wag her tail furiously as you take your hand away which should, in turn, induce a flurry of discharge if, in fact, she is in season.

If both of these fail to produce the desired effect the only alternative would be to take your nanny to a stud keeper who can board her, and leave her there until such time you're informed a satisfactory mating has taken place.

Occasionally, nanny might seem to come into season twice within a week. The first time will be purely a prelude to her coming into season properly about a week later. Billy will serve her on both occasions but the second service should be noted as the date of conception unless, as you check her three weeks later, the vulva is slightly swollen and there's discharge again, which means she hasn't taken or held so mating must be repeated. Once you're sure of conception date, five months later, (or 150 days), almost to the day, she will kid.

About 7 to 8 weeks before kidding's due, gradually increase her quota of concentrates (goat feed) to twice her normal ration, allowing her two handfuls night and morning. (8 ozs per day).

About 2 weeks before kidding's due, separate billy from her if he's running with her because nanny gives off an odour from about this time, not noticeable by humans but easily detectable by billy, which might incite him to mate with her and this could cause a stillbirth.

Nannies udder may begin to enlarge a few weeks before kidding but certainly with only one week to go, will enlarge considerably with the teats becoming prominent. At the same time the vulva will be seen to be swollen and there may be evidence of a slight discharge, as if she's in season.

A few days before kidding, a hollow either side of the backbone at the base of the tail will become evident. If nanny is housed together with other goats, now is the time to transfer her to an individual pen so that when the time comes she can kid peacefully without being disturbed by her inquisitive companions. She could, of course, kid unexpectedly during the night. Many's the time I've arrived at my goathouses to give them their breakfast and let them out, only to discover a nanny complete with fluffed up kids full of the joys of spring. But if you want to be with her (I like to be about keeping a watchful eye when kidding's imminent), observe her carefully a few days before kidding date for she may be a little early (or a few days late).

On the day of kidding the hollows either side of her back-bone by the tail will have become more pronounced. First signs that labour has started are restlessness and bleating on and off, maybe for a few hours and she won't venture far from her house. If you prefer, you can enclose her in her pen at this point. During this time remove the water bucket from her pen lest kid drops into it as it's born (this has been known) and make sure bedding is flattened so that kids don't become entangled in straw. It would perhaps be advisable to remove the water bucket from her pen last thing at night for the few days preceding kidding date in case she kids early and during the night.

It's rare for a nanny to kid outside unless she kids unexpectedly and has no access to her house, but if this happens take kid(s) and Mum into their pen as soon as possible. By gently picking up the kid(s) and holding them low so that Mum can see and smell them she will follow as you make your way to the pen. However, most usually, nanny will eventually disappear into her pen of her own volition and remain there. For the next hour or so she will paw at her bedding from time to time and repeatedly lie down and stand up again. A string of thick yellowish mucus will be seen hanging from the vulva before visible signs of straining begins. This last stage could continue for about twenty minutes or so before the kid is born. Sometimes the water bag breaks before the kid begins to emerge but often it appears first like a balloon emerging from the vulva within which the kid is encased and from which it will break out.

The normal sequence of delivery is that of the two front hooves appearing first followed by the nose, body and finally the two outstretched back legs. Any other way is considered abnormal and might cause complications so your vet should be called, unless it's something simple that can be dealt with yourself i.e. hooves bent backwards which can be corrected by carefully inserting a finger inside her and flicking the hooves forward. Before attempting this, hands must be clean and finger smeared with vaseline. If heavy straining continues for up to half an hour or so with nanny clearly becoming distressed, and no results, again, don't hesitate to call your vet. Any subsequent kids after the birth of the first, should arrive within the next half hour although in my experience second and third kids arrive one after the other very quickly.

Nanny may choose to lie down whilst kidding and as she stands up after the kid appears the umbilical cord will break. Alternatively, she may prefer to stand and in this case the cord will break as she turns to look at the kid as it gives its first yell. She will begin to clean it, ceaselessly licking and gently caressing and nudging the tiny face and body with her nose, all this loving attention interspersed with soft, deep throated little bleats of pride and pleasure in her newborn, a delightful scene to witness.

Nose and mouth must be cleared first to enable breathing, in fact, if she has more than one kid in a very short space of time and is turning from one to the other in a frenzy of activity, it won't hurt to give a helping hand by wiping around nose and mouth area yourself with a clean, damp cloth, allowing Mum to get on with the rest of the cleaning process. It might also be necessary to clear mucus from the inside of the mouth and throat with a clean finger.

Over the next two to three hours she will have completely cleaned and fluffed up the kid who, meanwhile, has been stumbling about searching for nanny's teats. In a strong kid this is sometimes accomplished within a matter of ten minutes or so but could take much longer for others, possibly up to three hours. In this case, I place the kid under Mum and squirt a little milk over its nose. This usually does the trick. The first milk the kid drinks is called the colostrum and is thick and creamy. It's very important that it has this as it provides immunity from infection and gives a healthy start to the kids' life.

If a kid lies apparently lifeless after the birth and no amount of nudging and coaxing by Mum can make it stand, first make sure nose, mouth and throat are clear of mucus then give it a brisk rubbing with a towel, this should stimulate it. If it remains limp, hold it up by the back legs and gently swing it back and forth between your legs. This often serves to revive an unresponsive kid.

The afterbirth (placenta) will drop within a few hours, never attempt to pull it away, it must be left to come away naturally. Nanny will probably eat it as part of her cleaning up routine but if not, it can be removed and buried or burnt. Remove any damp patches of straw and replace these areas with fresh straw. Ensure nanny has hay and plenty of fresh water to drink (I like to give warm water after kiddings), then leave her alone for the time being to rest and enjoy her offspring quietly.

Rejection of her kid by a nanny is rare but should this occur the first and most important thing to do is to get it dry and warm. A hair dryer is ideal for this purpose. After clearing mucus from nose and mouth and rubbing briskly with a towel, train the hair dryer over its body **making sure the air is not too hot.** Mission accomplished, the bottle feeding routine can be followed.

Do not give any goat feed for the first three meals after kidding. A plentiful supply of hay will suffice, then, starting with one handful twice daily, build up to her full quota of double ration (two handfuls twice daily), over the next few days. She will drink more water than usual whilst suckling a kid and continue with the increased quota of goat feed up until the kid has gone to its new home at the age of about twelve weeks, after which it can be reduced to the normal ration of one handful night and morning.

If your nanny should have triplets, increase her already doubled ration of goat feed whilst she's feeding them but keep an eye on the smallest of the trio for if it's getting little or no milk because of larger and stronger siblings emptying nannies udder before it can get a look in, you might have to bottle feed.

At first, the kids' droppings will be the colour and consistency of mustard then will change to a smaller version of mother's firm, dark, pellet like droppings.

For two or three weeks after kidding nanny will intermittently discharge a mixture of blood and mucus from the vagina. Sponging her down when necessary will keep flies at bay. The presence of a very distinct unpleasant smell during this time could indicate that part of the afterbirth has been retained or she has an infection, so your vet should be called to investigate. However, under normal circumstances, nanny will eventually settle down to the usual clear discharge every three weeks as she becomes regularly in season again.

Chapter ELEVEN   Arranged Colours

Cameroon kids are cute little replicas of their parents except that the coat could be a lighter or darker shade of grey than that of either the dam or sire.

Nigerian kids can be any colour other than partly or completely grey, or all white. Many attempts have been made by breeders to try to ensure that Nigerian kids be born of a specific colour by mating a nanny of the colour required with a self coloured billy. Reports that this always works have proved totally unreliable. One might be lucky and achieve the required result after one or two matings, but it is only luck and many other matings using the same breeding pair can, and do produce kids of quite different colouring to that of the parents. Other combinations of parent colourings have been tried but these too, fail to be consistent. Even triplets and twins are rarely identical in colouring.

I had a nanny who was white with black markings and one year when she was mated with my almost wholly black billy she produced triplets. One had black markings on white, the second white markings on black and the third was a tri-colour, mainly golden brown with markings of black and a little white. Another year a different nanny, dark chocolate brown in colour, mated with the same black billy also had triplets, one white with pale brown markings, another black on white and the third tri-coloured, mainly creamy beige with a little black and white.

Because male kids sell well as pets, I'm never so much interested in the sex of the newborn as I am in its colouring. Come spring, when kiddings begin, for me, there's always an element of wonder and surprise as each nanny, regardless of her colour (or that of the sire) produces offspring in a wide range of colours and markings. One thing's for sure, I'd never care to predict the colours of expected kids from any of my different coloured nannies.

Chapter TWELVE   Kid Rearing

Keep nanny and kid(s) shut in their pen for three or four days after the birth, this encourages the kid to keep suckling from her and will promote a good flow of milk. This also allows the kid to find its sea legs before venturing into the outside world. Although your other goats will have been sniffing the newborn kids through the rails of the pen, supervise their first outing because the adults will want to examine the young newcomer to the herd. This could be quite a daunting experience for the kid, who, in its bewilderment, might attempt to suckle from the nearest available teat (not necessarily Mum's) and as a consequence will find itself severely reprimanded by one very indignant goat who will butt the errant kid out of the way in no uncertain manner. But don't worry, the little one will soon be accepted into the herd and should it fall foul of a goat in truculent mood intent on proving its superiority, will quickly learn to skip out of the way.

Well within a week the kids will be poking their noses into Mum's feeding dish at mealtimes and at least getting the smell of the goat feed. In no time at all they'll again be copying mother by sucking at bits of hay. Long before it reaches the age of twelve weeks the kid will have been sampling water from the bucket and eating hay and goat feed along with grazing and at twelve weeks, although still enjoying the occasional drink from Mum, can be considered completely weaned.

As mentioned earlier, nanny's milk will dry up naturally, no assistance is necessary and although her udder will become very large and full, after a few days it will begin to diminish.

All kids must, of course, be wormed and I worm mine at about six to eight weeks of age.

Bottle feeding. There is a lamb teat on the market which has a plastic screw on base. This will fit on to a small coke bottle or a clear plastic fruit juice bottle having a screw cap. I find this type of teat ideal for the pygmy kid.

If you cannot obtain milk (including colostrum) from nanny you might know of a dairybreed goatkeeper, with a CAE free herd, who can supply you with both colostrum and milk, either fresh or frozen. Do make sure that the kid is fed some colostrum in the first ½ hour or so, if possible, and that it drinks this before any other milk otherwise the protection it gives will be lost.

Sterilize bottle and teat, fill the bottle with the quantity required, screw on the teat and stand in a small bowl or jug of hot water to warm through. Test the temperature on the back of your hand, it should be just warm.

It might take a little time and patience for the kid to accept this manner of feeding, don't forget the rubber teat isn't as soft and supple so hasn't the same feel in the mouth as Mum's teat. I don't want nanny to reject her kid so rather than taking the kid away from her to feed it, I take the bottle to the kid and sit down in the pen with kid and family. I like to have a clean, damp cloth to hand so that I can wipe around the mouth and chin after feeding, otherwise any dribbling will result in an unpleasant smell as the milk sours and congeals turning the hairs into stiff little spikes. I place the kid in a standing position between my outstretched legs, facing away from me with its rump pressed firmly against my tummy. I cup its chin in my left hand, pressing thumb and forefinger gently into the sides of the mouth and as it opens, quickly slide the teat in and squeeze the base of the teat so that it squirts in a little milk. A vigorous kid will soon get the idea and suck hungrily. You might have to resort to using a plastic syringe for the weaker kid, administering colostrum and milk in the same way as for liquid wormers. Give the kid a rest between each squirt. When it's stronger you can change over to the bottle and once it's used to the bottle you're home and dry, the kid accepting that it's fed differently to its siblings and looking forward to its regular feeds from the bottle yet still having the warmth, comfort and protection that only its mother can give, together with the companionship of its siblings. The name of the game is patience and perseverance and in no time at all the kid will be running towards you as you arrive and grabbing the teat in its mouth as you hold the bottle down towards it.

If you eventually have to change to milk from another dairy breed herd whose diet might differ from that of the first source, it's worth adding a spot of scouring mixture to the bottle of milk, twice daily on the first day and once on the second day, to prevent possible scouring. Droppings being normal with no mess around the tail will show that the kids digestive system has accepted the change in milk.

If you're unable to obtain goats milk from any source whatsoever, you'll have to resort to the powdered variety. Until fairly recently, a lamb replacement powder was considered best as a substitute for goats milk but there is now at least one product, "Caprilac", which is specially made for kids. To alleviate scouring, after mixing the powder with water as per instructions add a spot of the scouring mixture to the bottle daily for a few days.

The following table refers to a reasonably vigorous kid and must be adjusted to suit the needs of a tiny, weaker kid, when a "little and often" is preferable until it's sucking strongly. Encourage but do not force. The milk intake might be barely ½oz to 1oz after half an hours coaxing and occasional short squirts while the teat is in the mouth, between intermittent delicate sucks of the kid. In this case, the number of feeds over the day must be increased. As intake increases to an acceptable level so must you decrease the number of feeds until overall, in the first week, the kid is taking between 7½ozs to 12ozs spread over the day. If progress is slow you might enter into the second week yet the intake has only just reached the overall amount suitable per day, for the first week. But the kid will gradually catch up and the danger then is of overfeeding. For the kids own good don't go above the amounts shown week by week and the maximum intake of 20ozs spread over the day at weeks 5 and 6 should not be exceeded. The kid will probably suck strongly at the empty bottle as if it's still ravenous, but don't be tempted to give "just a little more".

Week 1.	5 or 6 feeds at regular intervals during the day of 1½ozs to 2ozs.
Week 2.	4 feeds of 3ozs spread over the day
Weeks 3 and 4.	3 feeds of 6ozs
Weeks 5 and 6.	2 feeds of 10ozs
Week 7.	2 feeds of 10ozs diluted with a little water.

Week 8.	2 feeds of 8ozs diluted with a little more water.
Week 9.	2 feeds of 6ozs (approx. ½ milk ½ water).
Week 10.	2 feeds of 6ozs (approx. 1/3 milk 2/3 water).
Week 11.	1 feed daily of 6ozs (approx. ¼ milk ¾ water).
Week 12.	1 feed of 4ozs nearly all water for 2 or 3 days.

Note:- If the kid is living separately from its mother, at 3 to 4 days begin to offer a little hay and concentrates, increasing gradually until by 11 to 12 weeks of age its consuming a full ration of 2ozs (one handful) night and morning (or 4ozs per day) together with ad lib hay.

Pygmy goats are noted for their hardiness and should kid easily without help but I must emphasize the importance of exercise for the in-kid female. The nanny who lies about in her pen all day and every day throughout her pregnancy, will, in all probability, have a difficult kidding which could result in a dead kid(s).

Do check fencing carefully. Adequate fencing for a fully grown pygmy goat may not be so for a kid who will quickly find any gaps possible for its small body to be squeezed through or under. Once through to the other side of the fence it will loudly and persistently call for its mother while unsuccessfully attempting to return to her, and what mayhem then as mother joins in with the urgent, penetrating bleats of her kid.

Don't part the kids from nanny before they're twelve weeks of age otherwise you could have a dejected, miserable Mum continuously calling and searching for her offspring, plus unhappy kids with worried new owners.

It's better for the kids if the buyer allows a few days for them to settle in, become accustomed to a strange environment and to get to know their new keepers before inviting friends and neighbours round to admire them.

On the day the buyers of my kids arrive to collect them, having already been wormed at 6 to 8 weeks of age, it only remains for me to powder them to ensure they'll be free of external parasites and trim hooves, barring one, which I leave so that I can demonstrate to the buyer how it's done. I also like to give them a little of my goat feed in a bag so that kids can be changed over gradually to a possibly, different goat mix, to prevent tummy upsets resulting in scouring.

I say my goodbyes and take a last look at the kids as I wave them off. If the buyer hasn't phoned me after a week has passed, I like to phone them, to check that all is well and that both buyer and kids are happy with each other.

Well, to cheer myself up after the last paragraph in which the events described are always a little sad for me at the time, I shall tell you a story about a faux pas I made in recent years, which might amuse you. It's just one of the milder ones of many I've made over the years and still make occasionally, despite my nineteen years experience, but then one never stops learning!

My daughter and three year old granddaughter were visiting me. The little girl was thrilled when I agreed to let her go down to the paddock to feed a kid I was having to bottle feed. It was the smallest of triplets and was getting very little milk from its mother because of the greediness of the two larger ones. I saw her off as she began the lengthy trek down to the paddock, walking slowly as I'd told her to do, hugging the bottle to her chest in case she dropped it. I called out, laughingly, "make sure you feed the right one" knowing, of course, that the kid concerned would run to her as soon as she passed through the gate, it being well used to this way of gaining sustenance. The other kids wouldn't even know what the bottle was for! Her Mummy was already in the paddock busy with camera taking snaps of the nannies and their offspring, nine kids in all that year, so she would be at hand to keep an eye on granddaughter.

Meanwhile, I was free to prepare the evening meal and in due course the little girl returned, flushed with pride and triumphantly handed me the empty bottle. A few weeks later, my daughter had her film developed and was showing me the prints. One of them was of granddaughter bottle feeding the kid, a lovely picture. But, I couldn't believe my eyes, indisputable evidence that she'd fed the wrong kid. Furthermore it wasn't even one of the triplets!

If there's a moral to this tale it must surely be "never underestimate the intelligence of the goat", but perhaps I should substitute "intelligence" with "greediness" in this instance!

USEFUL INFORMATION

Useful items to have available at all times:
    Antiseptic wound cream or powder.   Sharp knife
    Louse powder.   Plastic syringe
    Worming liquid or granules   Secateurs - straight bladed

Plus - if you intend breeding:
    Bottle   Teat   Vaseline
    Colostrum and goat's milk in freezer
    Scouring mixture   Small 2ml plastic syringe
    Scissors with rounded points

Neck measurements: Kids at 12 weeks old Approx. 10" to 11"
                             Fully grown nanny   "   13" to 14"
                             Fully grown billy   "   15" to 16"

Further Information

Pygmy Goat Club	British Goat Society
Karen Jephcott (Mem. Sec.)	34-36 Fore Street
Cedar Cottage	Bovey Tracey
Llandidloes Road	Nr. Newton Abbot
Penstrowed	Devon
Newton, Powys	TQ13 9AD

An increasing number of goat clubs are including pygmy classes in their shows. The address of your local goat club can usually be obtained from the British Goat Society or from your public library.

In East Anglia the two distinct types of Pygmy Goat are displayed and bred at:

Messrs. Smith & Pugh	Mr Trevor Walters
Maltings Rare Breeds	Park Farm
Daffy Green, Shipdham,	Snettisham, King's Lynn
Thetford, Norfolk IP25 7QQ	Norfolk PE31 7NQ

INDEX

	Page		Page
Addresses	49	Kidding	37 40 >
After Birth	*See Placenta*	Kid Rearing	44 >
Aggression	9		
		Legal Requirements	32
Bedding	22	Leading	27 34
'Billy Rag'	38	Lice	31
Bottle Feeding	45 >	Life Expectancy	29
Breeding	16 37 >		
" Twice Yearly	14	Males	16 >
Browsing	27	Maturity	7
		Milking	14
Cameroon Type	2 >	Movement (Legal)	32
Caprine Arthritis Encephalitis (CAE)	29	Neutering	*See Castration*
Castration	20	Nigerian Type	2 >
Collars	36		
Colostrum	41 45	Odour	16 38
Colours	43	Oestrus	14 37
Companions	33		38
Concentrates (feed)	26	Origins	2
Constipation	30		
Costs	27	Paddocks	23
Cross Breeding	17 18	Pigora cross breed	18
		Placenta	41 >
Description	2 >	Poisonous Plants	28
(Diagram)	6	Pregnancy	39
Dermatitis (Seborrhoeic)	31		
Disbudding	8 12	Registration (Legal)	32
		Rejection of Kids	42
Feeding	26 > 39	Runts	15
	42 46	Rut	16
Fencing	16 24		
		Scouring	30
Gestation	14 39	Season	*See Oestrus*
Grazing	7 28	Size	6 15
			49
Handling	16 27	Storage	25
	33 >		
Hay	26	Tethering	19 25
Hay Rack	23	Transport	32
Health	29 > 49		34
Height	6	Types	2 >
Hoof Trimming	30		
Housing	7 16	Water	27
	22 >	Wether	20
		Worming	29 47
Identification Marks	32		